12-11

MW00633101

DUE DATE

10-25-2022

BRODART, CO.

NATURE'S MAGIC

SHAPE SHIFTERS

WRITTEN BY
Q. L. Pearce

ILLUSTRATED BY
Marilee Niehaus

To Jonathan Adam Eisner
—Q.L.P.

To Lillian and Marvin, LaDonna and George
—M.T.N.

PRICE STERN SLOAN
Los Angeles

10 9 8 7 6 5 4 3 2 1   ISBN: 0-8431-2829-1

Designed by Heidi Frieder

THE PANGOLIN

A living pinecone

There are several different kinds of pangolins in Africa and Southeast Asia. Some live in
forests, and others live on grasslands. Some
make their homes on the ground, digging burrows with their long, sharp claws. Others take
to the trees, using their long tails to grasp
branches. But all pangolins have one thing in
common: Their bodies are covered with tough,
overlapping scales.

Pangolins are toothless anteaters. When roam-
ing at night they run the risk of being attacked
by hungry meat-eaters such as the ferocious
leopard. If there is no chance of escape, a
trapped pangolin quickly curls into a tight ball.
No matter where the predator prods or pokes, it is
met by sharp scales. The attacker will often aban-
don the prickly pangolin in search of easier prey.

THE PUFFERFISH

A little fish that's too big to swallow

The warm, tropical waters where pufferfish live are home to many other fish, too. Some of these are predators that find the slow-moving puffer the perfect size for a quick meal. The pufferfish, however, has a few surprises in store. While some kinds of these fish are smooth, many are covered with thornlike spines. By swallowing water (or air,

when it's near the surface of the water), the pufferfish can inflate its body like a balloon. As its skin stretches, the spines poke straight out, making the fish appear like a cross between a ball and a pin cushion.

Its expanded size and prickly covering are usually enough to make the pufferfish unappetizing to predators. Unfortunately for the small creature, those very features make it attractive to some humans. An air-filled pufferfish takes thirty minutes or more to deflate. Some fishermen capture the fish in this condition, dry them and sell them to tourists as curios.

THE PEACOCK

A bird with "eyes" on its tail

You may find it hard to believe that a peacock (a male peafowl) is a distant **relative** of the barnyard chicken. Native to Asia, this stunning bird now populates parks and public gardens all over the world. As with many birds, the male peafowl is more elaborate than the female (a peahen). The peacock is cloaked in feathers of shimmering blue and green and wears a dark crown of feathers on its head. Its tail, which may be up to five feet long, is made of about 150 iridescent feathers tipped with eyelike spots of purple, blue, yellow and bronze.

Usually, the peacock trails his magnificent tail behind him. However, during the breeding season he puts on an incredible display.

When he sees a peahen he wants to attract, the peacock raises his tail and spreads it in a gorgeous fan. He then struts back and forth, showing off his splendid plumage. It is easy to see where the saying "proud as a peacock" came from!

THE FRILLED LIZARD

A lizard that doesn't mind a stiff collar

The frilled lizard lives in the dense, dry forests and open grasslands of Australia. This reptile may reach thirty-two inches in length, but more than two-thirds of that is the creature's slim tail. When it is calmly hunting for beetles and spiders to eat, this animal looks like any other lizard. Folds of skin that lie flat against its neck are the only hint of this scaly tree-dweller's remarkable shape-shifting ability.

When startled by a hungry predator, the frilled lizard has a very different look. Instead of a lightweight lizard, the surprised enemy faces what appears to be a much larger foe. The frilled lizard has slender bones in its neck that are attached to a huge skin collar, or frill, that is up to seven inches wide. By extending the bones, it spreads the collar like opening an umbrella. Gaping its mouth open to show its tiny, needle-sharp teeth, the frilled lizard often convinces an attacker to look elsewhere for a meal.

THE THREE-BANDED ARMADILLO

A tiny tank

*A*rmadillo is a Spanish word meaning "little armored one." Covered with tough, bony plates that shield its back and sides, the three-banded armadillo of Brazil fits this description perfectly. Like other members of its family, this mammal (also known as an apar or a three-banded bolita) is equipped with long, sharp

claws for digging up meals of
worms, insects or roots.

Unlike its relatives, this armadillo doesn't appear to be an
exceptionally strong digger.
When faced with a predator, one
relative, the nine-banded armadillo, burrows into the ground
at lightning speed. The three-
banded armadillo, however, has
a different strategy. Even though
it has only three moveable bands
at the center of its armored back,
they are very flexible. At the first
sign of danger, the three-
banded armadillo rolls itself into a tight, round
ball. It stays in this
position until the
danger has passed.

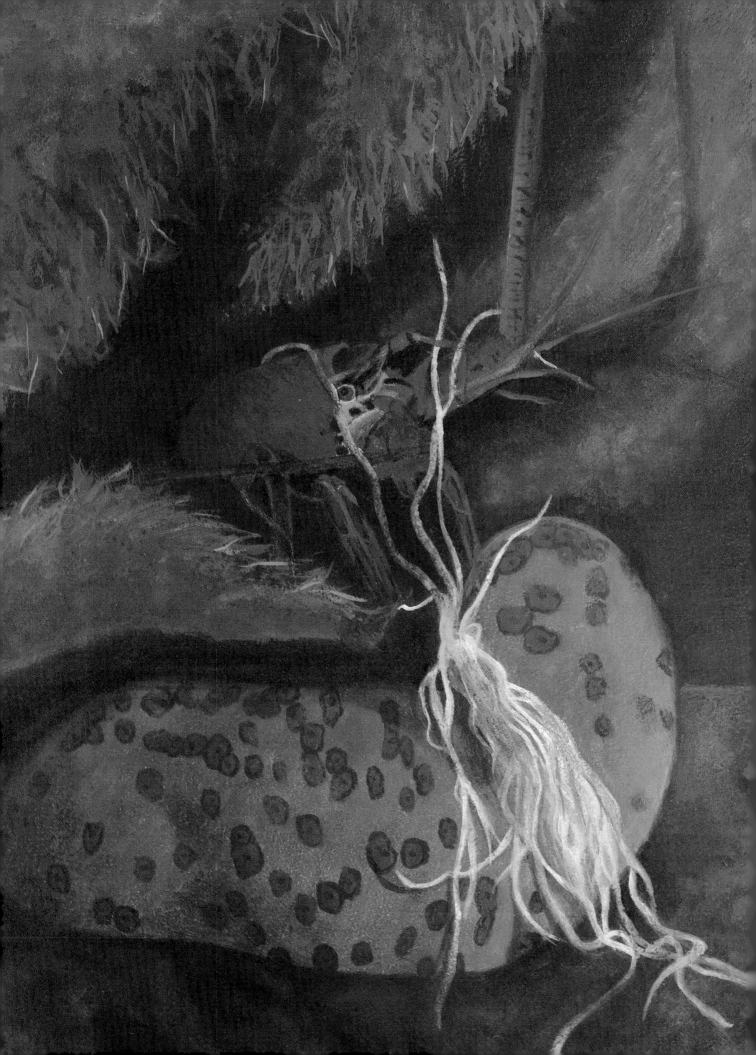

THE SEA CUCUMBER

A bagful of tricks

The rubbery-skinned sea cucumber looks a lot like a swollen version of the vegetable it is named after. Sea cucumbers, which can be from a few inches to more than a foot long, make their home on the seafloor. That can be a shallow reef or the very bottom of the ocean, five miles deep. Although it is poisonous, the sea cucumber occasionally may be attacked, perhaps by a lobster. When that happens the sea cucumber goes to pieces . . . truly! Its body breaks into several sections. Amazingly, the section containing the mouth and tentacles will grow into a new sea cucumber. That's not the only trick the sea cucumber has. It is sometimes plagued by a tiny fish called the pearlfish. The pearlfish swims in through the unfortunate sea cucumber's anal opening and feeds on its internal organs. The sea cucumber has no choice but to eject the little fish out the way it came—along with much of its own innards. Within just ten days the creature will grow new internal organs.

THE FLYING SQUIRREL

It's a bird, it's a plane, it's a . . . squirrel?

Except for an odd ruffle of skin along its sides, the flying squirrel doesn't look very different from most squirrels. Like other squirrels, it is sure-footed and agile, but that isn't the only talent it has. This extraordinary creature can actually glide from tree to tree for distances of 165 feet or more. That's about the length of three bowling alleys!

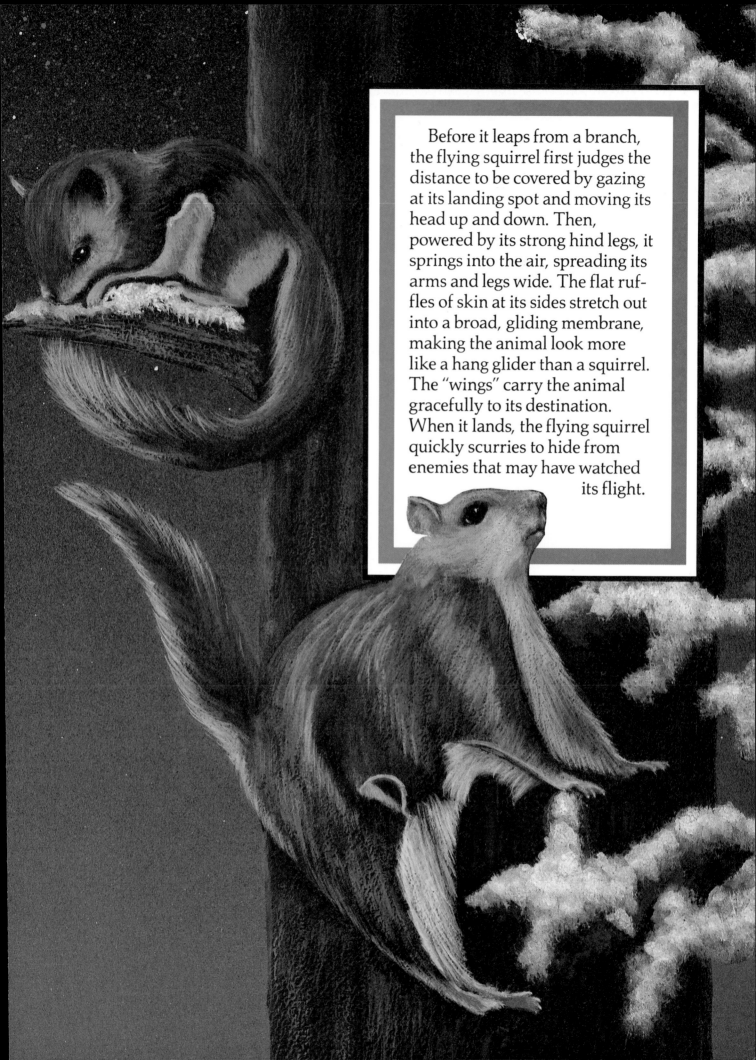

Before it leaps from a branch, the flying squirrel first judges the distance to be covered by gazing at its landing spot and moving its head up and down. Then, powered by its strong hind legs, it springs into the air, spreading its arms and legs wide. The flat ruffles of skin at its sides stretch out into a broad, gliding membrane, making the animal look more like a hang glider than a squirrel. The "wings" carry the animal gracefully to its destination. When it lands, the flying squirrel quickly scurries to hide from enemies that may have watched its flight.

THE MOSQUITO

An unwanted dinner guest

Usually it isn't difficult to tell the difference between a male and a female mosquito by what it eats. Both feed on plant juices and nectar. However, that is all that the male consumes. Most female mosquitoes, on the other hand, must drink at least one protein-rich meal of blood before her eggs can develop. Thus, the mosquito that gives you that annoying bite is a female. She has a slim, needlelike mouthpart that can pierce the skin. Through this she sucks up blood.

Generally, the female mosquito is a slender, dark, fragile-looking insect. But after dining on blood she takes on a very different appearance. Her abdomen (which stores the blood) can swell to more than twice its normal size as she drinks her fill. She flies away from her meal in a different shape than when she landed.

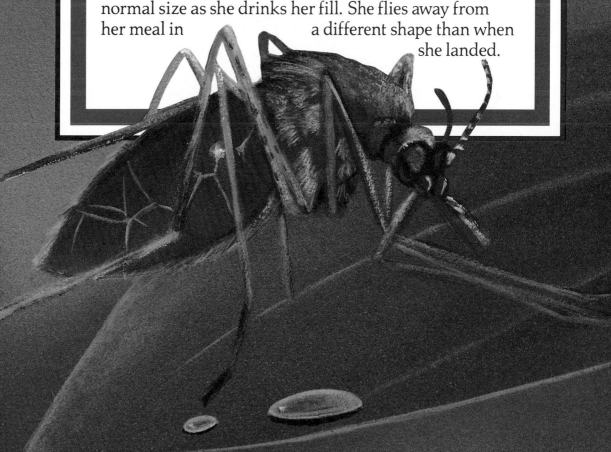

THE BULLFROG

A frog that goes a-courtin'

In spring, as night falls from Canada to Florida to Texas, the deepening darkness brings an amazing chorus. Along the muddy shores of ponds, lakes, and slow-running streams, the song of the male bullfrog fills the air. With an eight-inch-long body and ten-inch-long legs, the bullfrog is the largest North American frog. Generally, these chubby amphibians live

alone, but in spring males squat
near the edge of the water and
advertise for a mate.

Drawing air quickly into a
pouch in its throat, the frog in-
flates the stretchy skin like a
balloon. The expanded pouch
acts like a sound chamber to
enhance the call that the bullfrog
produces, making it loud and
deep. The call, a powerful
jug-o-rum, jug-o-rum, attracts
females and warns away rival
males. The patient male
calls over and over again
through the night until
he achieves his goal.

HONEY ANTS

Some insects spend their time just hanging around

Like most ants, honey ants of the southwestern United States and Mexico live in large colonies. Each colony is complete with a queen and hordes of busy workers foraging for food. The ants eat parts of plants and small insects, but they also eat a special food that is made by other insects. This sugary food, called honeydew, is made by aphids. The workers collect the honeydew and bring it back to the nest for storage. But how the honeydew is stored might surprise you.

Special ants, called repletes, are encouraged to eat as much honeydew as they can hold. That can be quite a lot, because the tiny replete's abdomen expands to nearly half an inch in diameter. After dining on honeydew, the repletes grow so large and heavy that they cannot move. They simply hang from the roof of their chamber like plump, living storage jars. Then, when food supplies are short, worker ants tap or stroke the repletes, encouraging them to regurgitate (spit up) the sugary fluid, one drop at a time.

THE DRACO LIZARD

A flying dragon

The rain forests of Indonesia are home to some amazing animals. Even so, would you expect to find a lizard that could fly? When at rest, the draco lizard may be overlooked. Its drab coloring blends well with the tree branches to which it clings. But if the lizard is threatened or spies an insect meal, a sudden change takes place. This small reptile fans out its ribs, spreads out flaps of stretchy skin into "wings" and glides through the air to another tree. In this way it can quickly escape danger or snap up the unsuspecting prey it finds there.

Also known as the flying dragon, the draco lizard usually swoops about 60 feet in a glide, but record distances are as great as 300 feet. The lizard has no need for camouflage in flight, so the undersides of its wings often have colorful patterns. Once it reaches its new perch, the draco lizard folds up its "flying cape" and fades into the background once again.

THE INDIAN COBRA

A snake that really sticks its neck out

The Indian cobra lives in jungles, rice fields, old buildings and anywhere else it can find shelter and plenty of mice, birds and frogs to eat.

When it is slithering along the ground, this slender, six-to-twelve-foot-long serpent with a short, flat head does not look very unusual. If it is annoyed or frightened, however, this normally peaceful snake takes on an unmistakable shape.

When startled by a careless human or threatened by an enemy such as the mongoose or tiger, the Indian cobra quickly rears up in self-defense. It can lift as much as half of its body completely off the ground. To seem larger and more frightening, the snake also flares the ribs along its neck, stretching the skin over them into a broad, flat hood. This makes its head appear to be up to four times its normal size. Most people heed this warning and leave the animal in peace. That's a wise decision, because the Indian cobra's poisonous bite can cause death in just fifteen minutes.

THE SOUTH AFRICAN ARMADILLO LIZARD

A lizard that gets all wound up

The veld or veldt is a Dutch name for the grassy plains and rocky open country of southern Africa. The scaly, ten-inch-long South African armadillo lizard makes its home here. Looking a little like a miniature, snub-nosed crocodile, the armadillo lizard is covered with sharp, spiny scales.

Unlike many lizards, this one doesn't have a "breakaway" tail. In fact, it uses

its tail to protect itself from predators. When an enemy approaches, the armadillo lizard scurries for cover, running under a rocky ledge or into a crevice. Then it grasps its tail with its front claws and puts part of it into its mouth. In this way, the armadillo lizard forms a defensive circle around its most vulnerable part—its soft belly. With dozens of prickly scales sticking out, the well-protected lizard looks more like a Christmas wreath than a reptile. When the predator loses interest and moves on, the armadillo lizard relaxes into its natural posture and shuffles out into the open once again.

THE MAGNIFICENT FRIGATE BIRD

An amazing expanding bird

Among the largest of seagoing birds, the nearly four-foot-long magnificent frigate bird has a wingspan of more than seven feet. An agile flier, this bird has been known to ride out a hurricane while on the wing.

The male magnificent frigate bird has striking jet-black plumage and a large ruby-red pouch of bare skin at its throat. This pouch is usually noteworthy only for its color. During the breeding season, however, the size and shape of the pouch play an important role. To attract a female, the male puts on quite a show by spreading his huge wings, shaking his head and crying out. He does this to draw attention to his gaudy pouch, which he can expand with air until it looks like a red balloon the size of a human head!

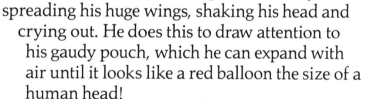

THE CRESTED PORCUPINE

Nature's pincushion

The crested porcupine of Africa is not an animal one would like to run into in the dark. Two and a half feet long, this member of the rodent family is armed with *thousands* of sharp quills on its back and tail. These spines are from twelve to sixteen inches long and make very effective weapons.

When the crested porcupine is left in peace as it searches for fruit, roots and bulbs to eat, its quills lie flat against its back. But when faced by a predator, this unusual creature turns its back, raises its hollow quills into attack position and rattles them in warning. If the enemy persists, the prickly rodent charges backward toward the unfortunate attacker, planting the spines in its paws or snout.

The porcupine does not shoot or throw its quills as some people think. The startled predator often races away, roaring in pain. If the animal is lucky, the quills can be worked out. But if any remain, they can cause a serious and sometimes deadly infection.